The Life of the Bumblebee

By the same author :

Bumblebees

The Life of the

Bumblebee

D. V. ALFORD BSc PhD

Northern Bee Books

The Life of the Bumblebee

© D. V. Alford BSc Phd

ISBN 978-1-904846-42-0

This volume was originally published in 1978
Reissued by Northern Bee Books in 2009

Published by Northern Bee Books
Scout Bottom Farm
Mytholmroyd
Hebden Bridge HX7 5JS (UK)

Contents

Preface

THE PURPOSE of this short book is to present an account of the life-history and habits of the bumblebee to the non-specialist reader. To this end, scientific terminology has been kept to a minimum and statements of fact have not been overburdened with examples. However, in so doing, I have endeavoured to avoid over-generalization and inaccuracy. References to scientific papers, reports and reviews have been omitted, and readers seeking further information on topics covered in this book should consult the more detailed monographs listed on page 75, the most up-to-date of which contain extensive bibliographies.

The bumblebee, or humble-bee as it was once known, has often inspired composers of music and verse, and is generally held in high esteem, being set apart from the layman's often ill-conceived dislike of so many insects. Fanciful tales have often been written about bees and other social insects, including ants, termites and wasps, yet truth is often stranger than fiction. It is hoped that this general account of bumble-bees will inspire a greater appreciation and understanding of these interesting and useful insects, and will convey something of the pleasures to be obtained from a study of their fascinating world.

D. V. A.
Cambridge

Chapter I

The features of the bumblebee

THE FLUFFY and colourful bumblebee is a well-known sign of friendliness, industry and thriftiness, being used in advertising to promote a range of commercial products or services from breakfast cereals to bank accounts. It is also very much a feature of the countryside, figuring as a symbol for nature conservation in many drawings or paintings, calendars and even postage stamps. The purposeful and inquisitive behaviour of the bumblebee has long attracted the attention and wonder of man, but relatively few people are acquainted with the complex private lives of these fascinating insects. This book is largely concerned with the habits of bumblebees. Before considering them in any detail, however, it is important to know the main features of insects, and bumblebees in particular, as this will contribute greatly to a general understanding and appreciation of the way that a bumblebee survives in a frequently hostile world.

Insects must be regarded as the most successful of all animals ever to have inhabited the earth, their success being due very much to their adaptability and generally small size. Today, there are about a million known species, and even now many thousands are still to be discovered or described. Insects are to be found in virtually all kinds of terrestrial and freshwater habitats. Only in the oceans and under extremes of climate are they unable to establish themselves.

Insects and their close relatives the Arachnids (spiders, harvestmen, scorpions, mites and ticks), Crustaceans (crabs, crayfish, lobsters, shrimps and woodlice) and Myriapods (centipedes and millipedes) are all Arthropods, a major and almost infinitely variable group of invertebrate animals. All are characterized by their often shell-like outer skeletons, their segmented bodies, and jointed limbs. Insects basically differ from other Arthropods in possessing only three pairs of legs and usually one or two pairs of wings.

Our most primitive insects, such as spring-tails and silver-fish, do not have wings and a few advanced types, including fleas, worker ants and certain beetles, have become wingless to suit their particular way of life. However, all insects share the same basic form of head, thorax and abdomen, with the outer body skeleton acting as an attachment point for the the muscles and forming a protective, and in parts flexible, skin not unlike a suit of armour.

The head is the main sensory and coordinating centre of the body, enclosing the brain and bearing the eyes, a pair of antennae or feelers which are sensitive to touch and smell, and the mouthparts. The form and function of the latter vary according to the habits of the insect. Aphids (blackflies, greenflies or plant lice), for instance, have hypodermic needle-like mouthparts used for sucking plant juices; mosquitoes have similar structures for sucking blood, while many beetles have powerful jaws adapted for biting and chewing. In a bumblebee the mouthparts are capable of performing various tasks including biting, lapping and lick-ing, their main features being a long, thin, hairy tongue (folded neatly away beneath the head when not in use), and a pair of hard mandibles or jaws. Secretions from various salivary glands, such as the mandibular glands, are also important for diluting food, scent-marking, and so on.

The thorax is the centre of locomotion, being essentially a box which supports the legs and wings, and packed with strong muscles. The form of the wings and legs of insects shows considerable variation and modification in different species. In the case of bumblebees, the wings are comparatively small, membranous, and more or less transparent; they lack the colour patterns developed in many insects, such as moths and butterflies. Insects typically possess small claws on their 'feet', and also small sucker-like pads, which enable them to gain a foothold on all but the most slippery of surfaces. Inevitably, the lighter the insect the easier it is to defy gravity, and in this context the bulky bumblebee is less well endowed than most. There is an important modification to the hind pair of legs of a female bumblebee, which is also seen in the even more specialized honey bee. This is the development of a basket-like arrangement of hairs to form the corbiculum, a collecting and carrying apparatus for pollen. As will be seen later, this structure, which is often

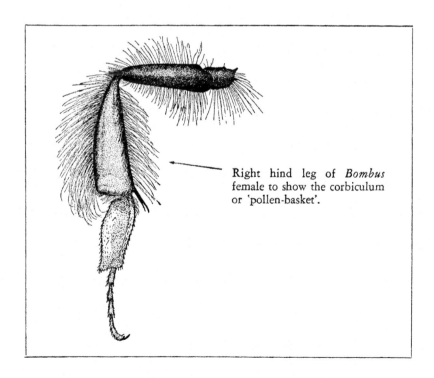

Right hind leg of *Bombus* female to show the corbiculum or 'pollen-basket'.

called the 'pollen-basket', is of considerable importance to the economy of the bumblebee.

The abdomen contains the digestive and reproductive systems of the insect. In a bumblebee it is formed by upper and lower series of overlapping plates and is joined to the thorax by a narrow, wasp-like waist. The gut is a long, convoluted tube divided into several distinct sections concerned with the storage and digestion of food, and the excretion of waste products. The foregut of the bumblebee is drawn into a large sac-like crop, or 'honey stomach', in which food is carried or stored before it is regurgitated or ingested. Eggs develop in a female bumblebee, like strings of enlarging and elongating beads, in eight long tubes called ovarioles. These tubes arise from a pair of ovaries which lead into a central vagina. A narrow duct runs from the top of the vagina into a small sack, known as the spermatheca, in which the sperms received during mating are stored until required. The reproductive system finally opens to the exterior through a single pore near the tip of the abdomen. Many insects possess a saw- or needle-like 'tube' through which eggs are laid. In bees and wasps, however, this structure has become modified and instead of performing a reproductive role, it now functions as a sting. The shaft of a bumblebee sting, like that of a wasp, is smooth and is able to penetrate and be withdrawn from a target, without harm to the user. This contrasts with the sting of a honey bee, which contains a series of barbs. These frequently cause the whole sting to be ripped from the tip of the bee's body, along with its associated poison-sac and muscles, and in such cases the hapless bee eventually dies. It should be pointed out that bumblebees are not aggressive insects, and they will normally only sting if they are themselves molested or if their nests are under attack. Male bees and wasps of all species do not possess stings and are, therefore, harmless.

Insects breathe through a series of finely branching tubules, the tracheoles, which spread throughout the body and contact all the main organs. The system opens to the outside through a series of small closeable holes, called spiracles, placed at intervals along the sides of the thorax and abdomen. In a bumblebee the system also contains large, thin-walled air-sacs which help to lighten the bulk of the insect for flight. The familiar buzzing of a bee is mainly caused by the passage of air through the thoracic spiracles and not, as is often supposed, by the movement of the wings.

An insect possesses an open blood system without veins and arteries, the more-or-less colourless blood spreading throughout the body where it bathes the various tissues and organs. Circulation of the blood is aided by muscular contractions and a flimsy, open-ended tube or 'heart' which runs along the top of the body cavity from the head to near the tip of the abdomen. Insects, unlike birds and mammals, are cold-blooded animals and their activities are, therefore, very much

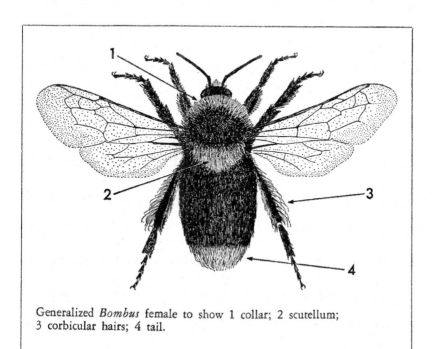

Generalized *Bombus* female to show 1 collar; 2 scutellum; 3 corbicular hairs; 4 tail.

influenced by the temperature of their surroundings. The bumblebee, however, is able to produce its own body heat, both chemically and by muscular movement akin to shivering and, when active, is more tolerant of cold conditions than many other insects.

The fat-body is another important internal organ of an insect. Most fat-body tissue is located in the abdomen, and it consists mainly of large cells which store fat and animal starch, often in considerable quantities. As will be seen later, these energy reserves are essential to the survival of bumblebees, especially during the winter.

A characteristic and endearing feature of a bumblebee is its long, furry coat which is formed by the body hairs. To a greater or lesser extent these hairs clothe the body and act as an important insulating layer. The 'skin' of a bumblebee is more or less black, the handsome and often striking appearance of these insects being entirely due to the colour of the hairs. Some kinds of bumblebee are very uniform in appearance but others show considerable variation, particularly in the proportion of black to yellow. The coat of a bumblebee is subject to considerable wear and tear, and is not replaceable. In old individuals the hairs are often faded or stained and some may be missing completely. Male bumblebees are often more brightly coloured than females, though this is not an infallible rule, but they are more generally recognizable by their slightly longer antennae, less pointed abdomens and absence of the pollen-collecting apparatus on their hind legs. Bumblebees rely almost entirely on flowering plants for food, and their very existence is dependent upon their obtaining adequate supplies of nectar and pollen. Towards this end, the hairs on a bumblebee are often finely branched or feathery, enabling pollen grains from flowers to adhere to them in large quantities. Using the legs as combs these pollen grains may be removed and, if required, either eaten or placed in the pollen-baskets for carriage back to the nest.

During its development an insect usually passes through an egg, larval and pupal stage. A bumblebee egg is sausage-shaped, pearly white in colour and about a couple of millimetres long. It hatches after a few days into a white, sluggish and hairless larva or grub with a distinct head and a fat, sack-like, distinctly segmented body. The larva is also blind, legless and wingless, and usually remains bent in the form of a 'C'. The mouthparts are simple but adequate for feeding on the pollen and nectar or honey supplied by the adult bees for its nourishment. As it grows, the larva must change its skin several times until it eventually ceases to feed and enters a brief, resting, pre-pupal stage. It then moults into the pupa, which is distinctly adult-like in form but still white and naked, with the wings merely represented by unexpanded pads. Close to maturity the pupa changes colour, eventually becoming black, and shortly afterwards the adult insect emerges. Once an adult, the body dimensions of an insect are fixed and these will not change, apart from any normal expansion or contraction the flexible body skeleton may allow. A small bumblebee is not a 'baby' and will not 'grow' into a larger one. Bumblebees, like ants, termites, honey bees and vespid wasps, are social insects, which means that they live in colonies sometimes termed 'insect societies'. It is in these colonies that eggs are laid and the young are reared.

Bumblebees occur in many parts of the world, and are most numerous in the temperate regions of Asia, Europe and North America. A few forms are able to exist within the Arctic Circle, including Alaska, Greenland and Siberia. Others inhabit the northern, coastal fringes of Africa, and some occur in Brazil. However, in the tropics and sub-tropics bumblebees are generally scarce and often absent or confined to mountainous regions. There are no native bumblebees in Australia but a small number of species, originally

B

introduced from Britain during the last century, are now well established in New Zealand.

There are many different kinds of bumblebee. In Britain alone there are some twenty-five known species, although only a few of these are particularly common (see Appendix). The 'true' bumblebees, with social habits, are placed in the genus *Bombus*. These have both males and females with the latter divided into two important castes, 'queen' and 'worker'. Queens are generally the largest individuals of the species and each is responsible for establishing and heading a bumblebee colony. Worker bumblebees are usually distinctly smaller than their mother queen. They perform various domestic tasks within the colony and also forage for food, but unlike queens, they do not mate and in the majority of cases their ovaries remain undeveloped. Structurally, however, queen and worker bumblebees are more or less identical.

In addition to *Bombus*, there is another major group of bumblebees known as 'cuckoo bumblebees'. These (genus *Psithyrus*) are similar to true bumblebees but they lack an industrious worker caste. Furthermore, they do not establish colonies of their own but invade those of *Bombus* and have their young reared by the *Bombus* workers. The intriguing habits and features of cuckoo bumblebees are described in detail later.

Chapter 2

Nest building and colony establishment

UNLIKE ANT, honey bee and termite colonies, those of bumblebees do not survive from year to year. Instead, they are established quite independently each spring by a new generation of bumblebee queens reared during the previous summer. These queens survive the winter by hibernating in the ground (Chapter 5) and finally venture forth when awakened by the warm days of spring. In particularly favourable areas, such as south-west England, the first bumblebees can emerge as early as February but if the spring is late, queens may not be seen in any numbers until April or even May. Also, it is well-known that some species leave their overwintering quarters before others. For example, *Bombus jonellus*, *Bombus pratorum* and *Bombus terrestris* are usually amongst the first individuals to emerge, while *Bombus humilis*, *Bombus lapidarius*, *Bombus pascuorum* and most kinds of *Psithyrus*, especially *Psithyrus rupestris*, tend to appear somewhat later.[1]

As soon as a bumblebee emerges from hibernation it is essential that she quickly finds a source of nectar, and during the early days of spring, bumblebees are a frequent and welcome sight as they gather food from various flowers. In areas where spring forage is scarce, bumblebees will find some difficulty in obtaining adequate nourishment and

---

[1] Brief descriptive notes on each of the British species of bumblebee, for which there are no generally acceptable common names, are given in the Appendix.

this can drastically affect populations later in the year.

Sallow or willow catkins, particularly the pussy or goat willow, are amongst the most important early producers of nectar and pollen for recently-overwintered queens. These catkins are also valuable to honey bees, solitary bees and many other useful insects, at a time when other suitable food sources are in short supply. In gardens, several spring flowers are extremely attractive to bumblebees, and in fine weather large numbers of queens may be seen collecting food from early-flowering heaths, heathers, aubrietias and bushes of flowering currant. Later, they will also visit fruit blossom and many other flowers.

Within a few weeks of emerging from hibernation, bumblebees begin to look for places where they can establish their colonies, and on sunny days they will spend many hours busily exploring neglected and overgrown gardens, rough banks, dykes and verges, woodlands, coppices, rubbish dumps, uncultivated fields, meadows and tussocky grassland, all of which are ideal nesting grounds.

In order to form her nest a bumblebee queen must usually locate a supply of suitable nesting material, such as dead leaves, fine grass, moss or hair, already collected together in a sheltered situation, either above or below the ground. Deserted nests of small mammals, particularly mice, shrews and voles, make ideal homes for bumblebees; less frequently, disused hedgehog nests, squirrels' dreys and birds' nests, including those built in birdboxes, are also utilized.

Some kinds of bumblebee will nearly always make their headquarters below ground level, often in mouse nests they have located in subterranean cavities. Such bumblebees are known as underground-nesting species; *Bombus lapidarius, Bombus lucorum* and *Bombus terrestris* are common examples. *Bombus lucorum* colonies are also frequently established in old mouse nests below garden sheds and other outbuildings. Underground nests are usually reached by way of existing cracks or other passages which, although often quite short, may sometimes be several metres long. Unlike wasps, bumblebees do not dig or enlarge their nest cavities and if space is short this will restrict future development of the colony. There are also several mainly surface-nesting species, including *Bombus muscorum, Bombus pascuorum* and *Bombus ruderarius*. They most commonly set up their colonies in deserted mouse nests found amongst heaps of rubbish or in the shelter of grass tussocks. Certain other bumblebees, including *Bombus hortorum* and *Bombus pratorum*, are very adaptable in their choice of a nesting place and may make their homes above or below ground level.

Land with a high water table is generally unsuitable for underground-nesting bumblebees and, in most instances, surface nests are also formed only in dry situations. One or two species, however, can tolerate damp conditions. Queens of *Bombus muscorum*, for example, which often inhabit marshlands and moorlands, will sometimes build their nests directly in deep moss.

Once a queen has found a suitable nesting place, which may take her anything from a few days to several weeks, she sets about making it habitable. Firstly, she manipulates the nest material with her legs and jaws, to form a tight-knit central mass about the size of a tennis ball. She then hollows out a cosy chamber which she lines with the finest of the nest fabric. Often much of this work has already been done by a previous occupant and the queen need only make minor alterations and repairs. At this stage, a bumblebee queen is very 'broody' and she will sit for long periods in the nest chamber, where heat from her body will soon rid the nest material of any dampness. Heat production is an important factor in the life of these insects and is achieved using nectar or honey as the basic source of energy. The temperature of an actively broody queen can reach about 38°C, although the body temperature of a non-active and non-broody individual will usually match that of her surroundings.

When leaving her nest for the first time, the queen carefully orientates to her surroundings, taking meticulous care to pin-point the exact whereabouts of both the nesting site and the nest or nest entrance. On subsequent flights she will make further, but less critical, observations until she is soon able to fly confidently to and fro without hesitation. Only in a strong wind or if the nest area has been otherwise disturbed, say by a fallen tree or passing animal, will she have difficulty in finding her nest. She is rarely lost for long, however, and will, if necessary, reorientate before taking her next flight.

On these first excursions, the queen visits flowers and will usually return home with a full load of nectar in her crop. Some of this is discharged onto the walls or floor of the central chamber and, as it dries, it helps to bind the innermost strands of nest material together. This also helps to improve nest insulation. Surplus nectar can also be used as food during extended periods of bad weather when the queen is prevented from leaving the nest.

By now the queen has well-developed ovaries, with several ripe eggs, and wax also begins to be discharged from special glands on her abdomen. She is then fully equipped to begin brood-rearing. Amongst her first tasks is the collection of pollen, which she carries back to her nest in the pollen-baskets on her hind legs. On her return she deposits the pollen onto the floor at the centre of the nest chamber and moulds it with her jaws into a cushion-shaped or pyramid-like lump. Next, she lays her first eggs which she places more or less vertically either around the outside or in small cavities within the pollen, precise details concerning the form and development of the egg clump varying somewhat from species to species. The egg clump is then covered by a layer of wax. This protective canopy is attached to the floor of the nest chamber and helps to anchor the brood clump. Egg-laying often begins before the pollen lump is completed although the full complement of eggs making up the first brood batch is usually deposited within a matter of a few days. In the case of less prolific species, such as *Bombus pascuorum*, the egg clump will normally contain eight eggs, which represents one from each of the queen's ovarioles. More prolific species, however, including *Bombus lucorum* and *Bombus terrestris*, may deposit up to sixteen eggs in their first brood batch. The completed egg clumps often have a characteristic shape. For instance, those of *Bombus pascuorum* and *Bombus pratorum* tend to be more or less cushion-shaped, while those of *Bombus hortorum* are wedge-shaped with a prominent horseshoe-shaped ridge at one end.

As well as an egg clump, the queen also constructs a waxen nectar pot about the size of a thimble. This is sited in the nest material at the entrance to the inner chamber and serves as an essential store for nectar that will be required to help sustain her during long periods in the nest. In some species, construction of the nectar pot is begun, if not completed, before a start is made on the brood clump.

Once the brood batch is established, the queen spends much of her time lying over the clump incubating the eggs. In most cases she faces directly towards the entrance of the brood chamber and by extending her tongue she can sip nectar from the nectar pot, as and when required, often without even having to leave her post. In this position, she is also well placed to defend her domicile against intruders.

On average, bumblebee eggs take about five days to hatch. The small larvae or grubs then immediately begin to feed on the lump of pollen provided for them. As this is soon exhausted, the queen is kept busy providing more and more food and, in order to satisfy the increasing appetites of her developing brood, she must continually forage for additional supplies of nectar and pollen. This is one of the most strenuous periods in the queen's life and poor weather conditions or shortage of suitable forage at this critical time can have a serious effect on the future success of the colony. Indeed, in unfavourable springs many potential bumblebee colonies will die out at this vulnerable stage of development.

As the larvae of the first brood batch grow, the clump expands both outwards and upwards, but by the addition of further wax the queen maintains the continuous, protective waxen canopy over the top. Also, with the increasing size of the larvae the position of individuals within the clump becomes visible externally with the appearance of slight, but distinct, swellings.

When in the nest, the queen spends much of her time feeding or incubating her brood and a definite groove is formed over the clump in line with the nectar pot and nest chamber entrance. The warmth given off by the queen's body is essential for development of the brood which, during incubation, is kept at about 30 to 32°C, heat being transferred to the clump via the more or less bare underside of her abdomen. The temperature of the brood-nest inevitably drops when the queen is away foraging for food but the insulating properties of the nest material keep heat loss to a minimum. Nevertheless, if for one reason or another the brood cools down excessively, development will be delayed, while nest temperatures below 10°C are likely to prove fatal. Particularly in cold conditions, therefore, it is important that the queen does not remain away from her nest for long.

Larvae feed in a common cell for about ten days but then, on reaching the fourth growth stage or 'instar', each spins a flimsy cocoon of silk and becomes separated from its neighbours. However, all still remain covered by the same wax canopy. At this point it is very evident that the larvae vary in size, those at the centre of the clump being distinctly larger than those at the sides; the centre-most individuals are also slightly more advanced in their development. The fourth larval instar is the period of most rapid growth. It lasts for about four days, during which time each larva consumes a lot of food and grows considerably. The larvae, having finished feeding, then spin more substantial, parchment-like cocoons in which they eventually pupate. The cocoons remain in close contact with one another to form a compact pupal clump. The queen then scrapes much of the wax off the cocoons and uses it to build a double row of new egg cells on top of the clump, along either side of the incubation groove. She lays a few eggs in each cell and continues to incubate the brood, lying stretched out along the, by now, very distinct and often highly polished incubation groove. The eggs in these cells will eventually give rise to the next series of brood batches that will develop upwards from the shoulders of the original pupal clump.

The pupal stage lasts for about two weeks, so that approximately five weeks after the first eggs were laid the young bumblebees are ready to emerge. Adults in the initial brood batch are all workers and much smaller than their mother; they appear over a period of about two to five days. In order to escape from the pupal cocoon a young adult must bite her way through the top, and she is usually helped by the mother queen or any of her sisters who might already have appeared. As soon as she is free, the young worker sheds any remaining portions of the sloughed off pupal skin and walks rather unsteadily to the nectar pot in order to drink. She then returns to the brood clump and settles down alongside her mother, much like a nestling chick or duckling. The body of a newly-emerged bumblebee is rather soft, while the wings are limp and curved downwards at their tips. Also, the coat is damp and silvery-grey in colour, with the hairs matted together and flattened against the body. The body

B*

and wings soon harden, however, the coat quickly dries out and, within a day or so, the bee acquires her characteristic adult coloration and is ready to assist the mother queen in the essential work of the colony.

Chapter 3

Social life and brood development

BEFORE THE APPEARANCE of the first workers in a colony, the foundress queen has probably already begun to feed her next batch of larvae, having toiled unaided for several weeks. However, with the eventual arrival of the young workers, which brings about the establishment of a social unit, her future tasks are greatly eased. There are few or no deaths in the first brood batch of a bumblebee colony so the initial worker force is usually composed of about eight individuals in less prolific species, and up to sixteen in the more productive, mirroring the number of eggs originally laid. An interesting feature is the size range of these bees. All are distinctly smaller than the foundress queen but they tend to fall into a series of larger individuals, which emerged from the central part of the brood clump, and several smaller bees from the outermost cocoons. In some cases, particularly if food during the development period was in short supply, the smallest individuals may be hardly bigger than house flies. The size differences amongst the first-brood workers determine the main workloads of individuals, the larger bees becoming foragers and the smaller ones household and nurse bees. Division of labour in a bumblebee colony is thus established from the outset.

In many cases, the foundress queen must continue to collect food, at least occasionally, even after the emergence of the first-brood workers, but as soon as the worker force in the colony is strong enough she will cease to forage and will remain full-time within the safety of the nest. It is then the responsibility of the workers to ensure that adequate food stores are maintained in the colony.

Bumblebees exhibit two distinct methods of feeding their young. The more advanced method is practised by bumblebees known as 'pollen-storers' (see Appendix) which, apart from the provisioning of the first egg clump and in a few more primitive species also priming later egg cells with pollen, provide all food for their larvae by regurgitating a liquid mixture of nectar or honey and pollen through temporary holes made in the waxen brood clump canopy. At the later stages of larval development, when they have spun their flimsy cocoons, food is then supplied through the sides where the silk and wax covering is often permanently open. A more primitive method, reminiscent of 'mass-provisioning' as seen in solitary bees, is adopted by the so-called 'pocket-makers'. Here, open 'pockets' are constructed beneath the brood clumps, usually up to three per clump, through which pollen moistened with nectar or honey is forced to form a basal cushion of food upon which the larvae feed. Food is then added to the brood clump from below throughout the larval feeding period. In the later stages of development the larvae of 'pocket-makers' are also fed by regurgitation in much the same way as those of 'pollen-storers'. Regurgitation is also the standard way of feeding their male and queen larvae.

Brood-rearing is not a haphazard affair but is carefully regulated to meet the varying needs and economy of the colony. The presence of pupal cocoons in a brood clump is usually the signal for the queen to build egg cells and lay more eggs. In this way, the brood is matched to the likely number of suitable workers that will be available to care for it. Some bumblebees, such as *Bombus lapidarius*, place their egg cells on top of the pupal cocoons but in such a way that they will not impede the emergence of the young bees. Other species, including *Bombus pascuorum*, site their egg cells on top of the comb where two or more pupal cocoons are joined together. The foundress queen constructs all her egg cells without the help of the workers. In building a cell she first makes a waxen foundation wall about 4 mm high. She then lays about six eggs, which are placed on the floor in a horizontal bundle. This takes but a few minutes and the queen then quickly roofs over the cell with more wax which she scrapes off the pupal cocoons. A few bumblebees, including *Bombus hortorum* and *Bombus ruderatus*, are termed 'pollen-primers' because of their habit of lining their egg cells with pollen. This behaviour is also shared by certain other 'pocket-makers', and also occurs in the most primitive members of the 'pollen-storer' group, including *Bombus pratorum*. Normally, however, food is not supplied until the eggs hatch.

Particularly in large colonies, considerable amounts of wax are required to keep the brood covered, most being secreted by the young workers. Wax is also required to build food storage vessels, and is continually re-used by the bees. Workers in populous underground nests often construct a thin waxen canopy over the comb, supporting it at intervals with pillars of wax. These covers are particularly well developed in colonies of *Bombus lapidarius*, which is a prolific wax producer, but are rarely, if ever, built by surface-nesting species.

Pupal cocoons tend to be more or less upright and, after the emergence of the young adults, those in suitable positions in the comb may serve as food storage cups. Before food is placed in them, however, they are cleaned out by the workers and, where necessary, repaired or enlarged with wax. These vessels normally contain only thin, recently gathered nectar that is quickly used up by the bees. Some species, particularly 'pollen-storers', also construct special waxen pots for long-term food storage, and these usually contain thick honey. Nectar, which is mainly a watery solution of sugars, especially sucrose, is converted to honey during storage following natural evaporation of much of the water, and enzyme activity which breaks down the bulk of the sucrose into the two simple sugars, fructose and glucose. Bumblebee honey usually contains about 20 per cent water, but is often very much stronger, in which case it is extremely thick and sticky. Honey bees deliberately evaporate water from their nectar by using the tongue, but this is not done by bumblebees.

Although pollen is sometimes stored in used or modified cocoons, 'pocket-makers' usually keep theirs in the pockets associated with the brood clumps. 'Pollen-storers', however, which often collect and store considerable quantities of pollen (hence their name), construct special waxen storage cylinders within the brood nest which sometimes tower well above the comb. A few 'pocket-makers' will extend their pollen pockets upwards into slight cylinders if large amounts of pollen are available, but this is comparatively unusual. Pollen is essential to both larval and adult bumblebees; it contains some fat and carbohydrate, but is mainly required as a source of protein.

Although vacated cocoons are often commandeered by the household bees for storing food, they are not re-used for brood-rearing. A bumblebee comb reflects this in its development, gradually enlarging upwards and outwards, leaving the remains of the oldest brood clumps at its base. Details of comb structure vary. Those of 'pollen-storers' tend to be rather loosely constructed, with individual brood clumps only clearly differentiated during the earliest stages of their development. This is particularly noticeable in nests of *Bombus lucorum* and *Bombus terrestris* where, apart from those in the very first brood clump, the pupal cocoons are often only superficially attached to their immediate neighbours. Such combs have little apparent organization. On the other hand, the comb in a *Bombus lapidarius* colony is usually particularly well structured, and is one of the most attractive of all our species. Combs of several 'pocket-makers', including *Bombus hortorum* and *Bombus subterraneus*, often appear somewhat disorganized but others show a very characteristic form. Those of *Bombus ruderarius*, for example, are composed of several overlapping tiers with the youngest brood clumps at the top, while the comb in a *Bombus pascuorum* colony tends to be divided into an outer ring of younger brood clumps surrounding several older clumps, long since vacated. In such colonies it is often possible to find the remains of the original pupal clump at the nucleus of the comb.

As time passes, the number of workers reared in the colony increases considerably. The largest bumblebee colonies are found in the tropics where worker forces of over 2,000 individuals have been reported. In Britain, the most successful nests will contain a few hundred workers, but most species produce far fewer than this. In general, underground nests tend to be more populous than those of surface-nesting species. The smallest colonies are found in the Arctic, where the period of bumblebee activity is extremely short. *Bombus polaris*, for instance, produces only one brood of workers before rearing males and new queens.

In a well-established colony, a pecking order exists amongst the workers, with the most dominant and aggressive of them showing some degree of ovarian development. Usually, the dominance of the queen prevents these workers from laying eggs, but as the season advances her reign becomes less secure. In normal circumstances workers do not succeed in their attempts to establish their own brood and even if they do manage to lay some eggs, the queen manages to destroy them. However, if a foundress queen dies prematurely, or becomes in some way incapacitated, a small number of the most dominant workers will take her place but, being unfertilized, their eggs will only give rise to male brood. In some species, notably *Bombus lapidarius*, the most aggressive workers frequently attack the queen's youngest egg cells and attempt to eat the contents. The queen defends this brood as well as she can, but while she is in conflict with one worker a cell may be raided by another. This apparent breakdown in the bumblebee society normally occurs at about the time of peak worker numbers, but is not as suicidal as it appears, and serves a useful purpose by regulating brood numbers in preparation for the establishment of the sexual brood.

The body size of workers in a colony shows considerable variation. This is particularly noticeable in 'pocket-makers' where the method of brood-rearing is more haphazard and competition for food by the larvae in a brood batch is considerable. At any given time, as in the first batch of workers, the largest individuals tend to be the foragers and the smallest the household bees, but the system is not a rigid one and, if colony needs demand it, established foragers will perform various domestic duties. Also, the household bees act as a foraging reserve so that, if and when necessary, the number of field bees may be augmented or, in the event of casualties, replaced. Such flexibility is of considerable advantage in the case of bumblebee colonies since, unlike most other social insect societies, adult numbers are relatively small. Foraging is the most dangerous occupation and, in summer, many foragers will die when no more than two to three weeks old. However, the average life of a bumblebee worker is about six weeks, with some of the most established household bees surviving for several months.

In populous nests, several of the workers function as guard bees and station themselves at the nest entrance to defend the colony against potential enemies. Foragers returning to the nest are also scrutinized. Each member of a bumblebee colony acquires a characteristic body odour from the comb, food and nest material, which makes them instantly recognizable, and any strangers attempting to gain access to the nest will be detected immediately and probably expelled. It is noticeable that guard bees are less aggressive towards strangers of their own species, while they also seem incapable of defending their nests against certain parasites (Chapter 7).

External disturbance of a bumblebee nest will usually send the occupants into a brief frenzy of activity and a loud buzzing will be audible for several seconds. Continued interference will result in some of the bees leaving the colony to investigate. Some species, such as *Bombus pratorum*, are very mild-tempered, even under provocation, but at the other extreme *Bombus terrestris* and *Bombus muscorum* are particularly tenacious in defending their homes and will mount persistent attacks against invaders, including man. Most bumblebees fight with the sting and jaws, but a few species ward off attacks by coating intruders with honey. Such bumblebees have been called 'honey-daubers'.

Throughout the life of a colony the workers must maintain a suitable breeding environment within the nest. Brood-nest temperatures are kept at about 25 to 30°C, but optimum conditions vary from species to species. If colonies become too hot, then the bees fan with their wings to increase ventilation. They may also make temporary holes in any wax canopy over the comb or in the surrounding nest material. Conditions within the colony are particularly stable when the worker population is at its peak and this coincides with the important period of male and queen production. Humidity within the nest is also carefully controlled to prevent excessive drying of the brood and dampness which might increase the development of moulds and onset of diseases.

When a queen lays eggs they are normally fertilized by sperms stored in her body since the time of mating in the previous summer. Such eggs will give rise to females and whether these will be queens or workers is determined during the larval stage. If eggs are not fertilized then, as with those laid by workers, they only develop into males.

When worker density in the nest is at a higher level, usually after the production of several batches of worker brood, the queen is triggered into laying her first unfertilized eggs, thereby initiating the sexual brood. Male larvae consume much less food than females and so at this time, with worker activity at its height, the economy of the colony is considerably enhanced. Unless poor weather intervenes, it usually follows that food reserves in the nest then become extremely plentiful. There may be several batches of male brood before the queen returns to laying fertilized eggs. Larvae hatching from the latter are then given abundant food and, instead of becoming workers, they will develop into young queens. Details of the mechanisms which determine whether a bumblebee larva will become a worker or a queen are complex and vary from species to species. However, basically, a female larva is caste plastic (capable of becoming either a worker or a queen) up to a certain stage in its development and it is quantity of food received and consumed up to that point that is all important. Apparently, the presence of male pupae in a colony stimulates the nurse bees into paying particular care and attention to the young female larvae, thereby ensuring that they become queens. If the number of workers or amount of food available is insufficient to ensure this, then the brood will be culled to a suitable level, often by egg-eating. It is interesting that once sexual brood is initiated, colonies no longer return to the production of workers. Exceptions occur in a few species, such as *Bombus jonellus* and *Bombus pratorum*, where one or two of the young queens may themselves initiate brood-rearing instead of entering hibernation. However, second-generation colonies are uncommon and confined to those species which complete their developmental cycle rather early in the year.

The duration of colony development varies considerably. Many will die out for a variety of reasons before producing males, and even fewer will successfully rear young queens. However, a large and successful colony of a prolific species such as *Bombus terrestris* may raise several hundred sexual forms and there are usually twice as many males as queens. Some species, including *Bombus jonellus and Bombus pratorum*, have a short colony life-cycle and complete their normal term by early summer; even large nests of *Bombus terrestris* will normally have died out before the end of August. Those of late-establishing species, such as *Bombus humilis* and *Bombus sylvarum*, however, often survive until the autumn. The same is true of *Bombus pascuorum* colonies which, although initiated in April or early May, often persist throughout the summer and well into the autumn.

Male bumblebees do not remain in the nest for long and usually leave for good on their first flight, when but a few days old. Their subsequent life, away from the nest, is described in the next chapter. The young queens, however, stay with their maternal colony for anything from one to several weeks. Here they feed on the food stores and quickly develop large fat-bodies whose contents will later sustain them over the long winter months. Only if food reserves in the colony are low do the young queens forage, although on sunny days they will fly out of the nest in order to find a mate. Soon, however, when fat-body development is complete and mating has occurred, they abandon the nest in order to take up their overwintering quarters. By now, brood-rearing will have ceased and any remaining workers, along with the old foundress queen, if she still survives, soon die. The bumblebee colony, its many tasks completed, is then at an end.

Chapter 4

Field activities in summer

ALMOST ALL THE FOOD required by bumblebees is collected from flowers in the form of nectar or pollen and, during the summer months, bumblebees will work very long hours, often foraging from dawn to dusk and sometimes even longer. Foraging usually reaches its peak by mid-morning but in hot weather there is a distinct lull in activity during the middle of the day. Bad weather, such as heavy rain or fog, will also influence flight activity, although bumblebees are renowned for their ability to continue working in conditions that seriously hamper or totally prevent other insects from flying. Bumblebees are particularly well adapted to withstand and continue operating under cold conditions; in fact, they can if necessary fly at temperatures below freezing, and in snow. Their hairy coats, for example, are excellent insulators which help them retain body heat. In particular harsh areas, such as the Arctic, their coats are rather long and thick, and the hairs tend to be dark in colour to help the bee take full advantage of solar radiation.

Heat production by bumblebees has been mentioned in connection with broodiness and brood incubation. It is also essential for flight. Indeed, unless a bumblebee can raise its thoracic temperature to about 30°C or more it will be unable to become or to stay airborne, as she will be too sluggish and the wings will beat too slowly. An actively-flying individual will normally maintain a thoracic temperature of 35–

40°C and before take-off will undergo a period of warm-up, which may or may not involve muscular activity. The key to a bumblebee's success as an all-weather flier lies in her ability to raise the body temperature by enzyme activity within the flight muscles, which breaks down certain sugars and releases energy in the form of heat. Apparently, the particular enzyme involved is not present in the muscles of either cuckoo bumblebees or honey bees.

Generating heat is expensive in terms of fuel, and a foraging bumblebee will therefore allow its body temperature to drop when she settles on a flower. If food is being collected from compact inflorescences, such as buddleia, dandelions and sunflowers, the bee can walk from one floret to another and the thoracic temperature will drop below the threshold for flight with a corresponding saving in energy; warm-up then occurs before the bee takes off again. When foraging on isolated flowers or scattered inflorescences, however, body temperature is usually kept just above the threshold for flight, as the inter-flight periods are comparatively short.

Light is also a limiting factor for flying and, in summer, large numbers of workers will remain away from their colonies overnight, having been caught too far from home as darkness fell. Such workers often sleep on flowerheads, and in the morning may be found still torpid and sometimes even coated with dew. If disturbed, they may as a warning raise one of their mid-legs in the air and produce a low-pitched buzz. They may also threaten to sting but will be too sluggish to fly away until they have warmed up sufficiently.

Although bumblebees tend to forage close to their nests they do not fly about indiscriminately, and individuals will consistently collect food from particular forage areas. Scarcity of suitable food sources may require longer foraging trips than normal, perhaps up to a kilometre or more, but the further a bee has to travel the more food she will consume as fuel in doing so and the less economic the journey will become. In general, larger foragers tend to collect food over greater distances than smaller ones and, as they can carry larger burdens, they tend to make fewer trips per day; they also visit more flowers on each journey.

Foragers are able to navigate by the sun, and will take account of its changing position during the day. As in the case of the foundress queen at the time of nest establishment, they also perform orientation flights on their first few field excursions and are soon well acquainted with the position of their nest and its surroundings. Foragers also memorize the position of particular forage sites. Young queens orientate only casually to the maternal nest, while the males, which usually leave the nest for good on their first flight and then live independently in the field, do not orientate at all.

Flowers visited by insects are often very colourful and large in size. However, if small, they are frequently formed into compact clusters to make them more conspicuous. Many flowers are sweet-scented, which also helps to attract insects which may be necessary to pollinate them. Bumblebees have an acute sense of smell and can even detect the fragrance of flowers, such as viper's bugloss and yellow toadflax, considered odourless by man. Flowers have many adaptations which help or encourage the right kinds of insects to visit them. Foxgloves, for instance, have spotted patterns called nectar guides which help direct novice foragers towards the nectar-secreting nectaries at the base of the corolla tube. Bumblebees are especially attracted to irregular flowers, particularly those displaying a three-dimensional picture; many such flowers, including snap-dragons, can only be entered and pollinated by large-bodied insects. Indeed, pollination of some plants, including heartsease, is almost entirely achieved by bumblebees.

Although bumblebees visit many different flowers, individual species do have certain preferences, as can easily be seen by following them at work on wild or garden flowers. Irrespective of differences in their body size, bumblebees can be sub-divided into 'long-tongued' and 'short-tongued' species, tongue length having a considerable influence on a bee's foraging behaviour. Long-tongued species, such as *Bombus hortorum* and *Bombus ruderatus*, are specialized feeders and they restrict their foraging activities to flowers with long corolla tubes, many of which have nectaries so far from the mouth of the flowers that other bumblebees cannot normally reach them. However, short-tongued bumblebees have perfected a method of stealing nectar from such flowers by biting holes in the back of the corolla tube close to the nectaries. Hole-biting by bumblebees, particularly *Bombus lucorum* and *Bombus terrestris*, is often seen in the case of flowers such as beans, clover and comfrey. Bumblebees which damage flowers in this way are known as 'primary robbers', and it has been calculated that a single forager in a runner bean plantation will perforate about 2,000 flowers in a day. Bees which do not themselves bite holes, but have adopted the lazy habit of collecting nectar through those cut by robber bumblebees, are known as 'secondary robbers'. On bean crops, for instance, the activities of robber bumblebees will have a noticeable effect on the foraging habits of nectar-gathering, long-tongued bumblebees, and honey bees, which will quickly learn how to rob the flowers. Pollen-gatherers however, will continue to enter flowers as before. It should be emphasized that hole-biting is not in itself detrimental, and perforated flowers are still able to be pollinated and will set seed if visited by insects in the normal way.

Short-tongued bumblebees are opportunist foragers and will visit all manner of food sources, including honeydew secreted by aphids or psyllids (jumping plant lice) and even the sweet sap issuing from wounded trees. Nevertheless, they will show definite preferences for both nectar and pollen collection. Pollen-gatherers are sometimes influenced in their choice of flowers by the smell of the predominant species of pollen in the colony's food stores.

Pollen readily adheres to the hairy coat of a bumblebee, and at intervals the grains are combed out by the legs and either discarded or placed in the pollen-baskets for carriage back to the nest. The amount of food collected by any one individual varies considerably, but can be as much as half her own body weight. It follows, therefore, that larger workers are the most effective foragers. They also tend to be pollen-gatherers rather than nectar-gatherers. This division of labour, however, is not a strict one and workers of all sizes will vary their foraging habits according to the particular requirements of the colony; also, on any one trip, an individual may collect both nectar and pollen.

In summer, large numbers of dead bumblebees are often found beneath lime trees, and it is sometimes thought that pesticides are to blame. However, nectar secreted by the blossom is itself intoxicating and often fatal to bees.[1] Workers and males of *Bombus lucorum* and *Bombus terrestris*, which are much attracted to lime, are the most frequent victims. Drunken specimens around the trees are commonly attacked by wasps and birds, especially great tits, while dead bodies on the ground are often plundered by ants. Foraging is a hazardous occupation and individual workers usually survive for only a few weeks. Many will be killed by predators or parasites (Chapter 7) and numbers will be constantly diminished by various accidents. Until the rearing of sexual brood, these losses are off-set by new recruits but, by the end of the summer, numbers of most species will have dropped considerably and, in some cases, foraging will have ceased altogether.

---

[1] Nectar of *Tilia petiolaris* is certainly toxic to bees; other limes, however, including *Tilia olivieri*, are probably safe.

During the summer months, young queens are rarely noticed on flowers, although large numbers of male bumblebees, like workers, may be seen extracting nectar from various flowers. In poor weather conditions, and at night, males will often sleep or doze on flower heads such as knapweeds and thistles. Their main preoccupation, however, is to find and successfully mate with a young queen of the same species as themselves.

In some cases, male bumblebees collect around nest entrances in the hope of discovering a suitable partner and, if visual contact is made, they will chase after their quarry and perhaps even pursue her into the colony. In these circumstances pairing may take place within the nest, but although such behaviour is sometimes adopted by *Bombus ruderarius* and *Bombus subterraneus*, it is by no means typical of the majority of our species. Most bumblebees will in fact mate in the field, and the males have developed a complex pattern of behaviour to attract and enable them to locate suitable females.

On any sunny day in the summer, bumblebee males will establish what are known as 'flight-paths'. In essence, these are circuits around which they fly in seemingly endless processions, pausing at intervals in particular places, such as the stump or base of a tree, which they have previously scent-marked with a secretion from their mandibular glands. Males calling at these visiting places are sometimes very numerous and are often mistaken for workers entering or leaving their nests. Bumblebee flight-path activity has fascinated many observers and is common to both *Bombus* and *Psithyrus*.

Precise details of the circuits and visiting places vary from species to species. For example, males of *Bombus ruderatus* chase about at the tops of tall trees, but flight-paths of *Bombus hortorum*, which is one of the most frequently observed species in tree-lined habitats, are usually established near to ground level. Here, the same visiting places, such as dark recesses at the bases of trees, are often selected by the males year after year. Males of many bumblebees avoid trees altogether and set up their territories in grassy or at best bushy areas. *Psithyrus campestris* and *Psithyrus rupestris* males, for instance, often fly rapidly over sloping grassland, the latter species favouring areas of short turf with a north-west aspect. Separation of the reproductive activities of the various species has obvious advantages and even when two kinds share similar flight-paths, as do the closely related *Bombus lucorum* and *Bombus terrestris* which both often fly at tree-top height, peak activity for each species occurs at a different time of the day.

Visiting places are usually scent-marked at the start of each day's activities, and perhaps also after a heavy shower of rain, each species producing a different scent to which only bees of their own kind will be attracted. These scents, of which some are very pleasant, are readily detected by man. If a young queen is enticed to a visiting place she is soon found by a male who immediately pounces upon her, usually knocking her to the ground in his attempt to copulate. A pair of bumblebees may remain united for anything up to an hour or more, but to avoid disturbance from other passing males they do not remain at the visiting place for long. Although pairing bumblebees are occasionally seen flying in tandem, they are more often found resting on the ground or on the foliage of trees and bushes. Most females will mate only once before they hibernate, but a few species are known to be promiscuous.

The flight-path phenomenon of bumblebees has probably evolved from the same kind of system still seen in several solitary bees, where the males scent-mark flowers and then fly about close-by in the hope of meeting foraging, nubile females. With the development of a worker caste, the spatial separation of foraging and reproductive activities is clearly advantageous to bumblebees, as this overcomes any difficulties the males experience in distinguishing young queens from workers, and allows the latter to forage unmolested.

With the ending of summer, bumblebee workers become less and less numerous and the rapidly declining numbers of males spend a far greater proportion of their time away from their flight-paths, lazing on flowers or merely basking in sunshine. By autumn, colonies of most species will have produced their young queens and, perhaps, already have died out. However, nests of some kinds, such as *Bombus pascuorum*, which give rise to sexual forms somewhat later in the season than most, may still show signs of life throughout September and even into October. Once autumn sets in, however, bumblebee activity is soon ended and the first night frosts will dispose of any late stragglers, leaving only the young, hibernating queens as custodians of their species.

Chapter 5

Bumblebees in winter

As far as bumblebees are concerned, winter is very much a time of inactivity. Nevertheless, it is an extremely important period in the life-cycle of these insects. With the decline of the colony in late summer or autumn, and the eventual death of the old foundress queen, workers and males, survival of the species from one year to the next is entirely the responsibility of the young, mated queens. It is these individuals which then pass the winter in a state of hibernation. In Arctic regions, where the season for bumblebee activity is measurable in weeks rather than in months, the period of rest undertaken by the young queens is necessarily very protracted. Even in temperate countries, however, including the British Isles, some bumblebees may remain inactive for up to eight or nine months. In sub-tropical or tropical areas, such as Brazil, bumblebees do not need to hibernate, although the young queens may spend the unfavourable dry season in their maternal nests; indeed, in such areas, whole colonies may survive for more than a year. But these are exceptional examples, and the pattern of life established under temperate conditions is more typical.

The hibernation period is often assumed to begin in the autumn, as days get shorter and temperatures lower. However, young bumblebee queens do not normally wait around until the autumn before disappearing and will, in fact, take up their winter quarters as soon as they are in a fit state to do so. The young queens of early-nesting species, such as *Bombus jonellus* and *Bombus pratorum,* may even begin their long winter sleep as early as June.

Bumblebees do not normally hibernate in their nests but instead they will select special places elsewhere in the field which, hopefully, will provide the right sort of conditions and adequate shelter to guarantee their survival through to the following spring. In detail, the kinds of overwintering situations chosen by bumblebees vary from species to species. However, most queens will dig themselves into the soil, often selecting moss-covered banks, the sides of dykes or ditches, and other suitable stretches of shaded, sloping or undulating ground. Young queens of the common *Bombus lapidarius,* for instance, may often be seen during the summer months exploring or digging their way into such places. Several kinds, including *Bombus lucorum, Bombus pratorum* and *Bombus terrestris,* will often seek shelter close to the trunks of trees where they will settle down in the soil immediately below a layer of moss or leaf litter, almost invariably on sloping, shaded ground; others, including the widespread and often abundant species *Bombus pascuorum,* will hide away in the ground below a sheltering layer of herbage.

Overwintering sites suitable for bumblebees are necessarily well-drained to minimize the danger of waterlogging during the winter and to prevent any hibernating queens from becoming infected by diseases, which could develop if conditions were too moist. Excessively wet conditions would also impair a bumblebee's ability to withstand cold conditions. On the other hand, the soil in which a bumblebee is hibernating must not be too dry as this could lead to the death of the queen through desiccation; in fact, a bumblebee seeking to find a suitable place to overwinter will only elect to dig into a damp soil. Shelter from the direct rays of the sun, both before and after leaf-fall, is another essential requirement for an overwintering site. Shade will help prevent the ground from drying out and will ensure that soil temperatures are kept down and that they will not fluctuate excessively. If the ground close to a hibernating bumblebee were to become bathed in sunshine, soil temperatures would inevitably rise and this could lead to disturbance, premature emergence or even death of the queen. To avoid this, hibernation sites most frequently have a north or north-west exposure.

It was once believed that bumblebees overwintered 'deep in the bowels of the earth' so that they would escape the severest of winter conditions. In fact, they remain quite close to the surface, often settling down to sleep at depths of no more than five to fifteen centimetres. There is no necessity for a hibernating bumblebee to burrow below the frost line since those living in areas where the ground is likely to become frozen will manufacture their own antifreeze in the form of glycerol. This antifreeze will help prevent the formation of ice crystals in their body tissues and, at least in some species, will give protection at temperatures well below freezing. Cold-hardiness of bumblebees is particularly well seen in those species living in areas subject to very cold winters. In addition, some bumblebees will habitually over-winter at slightly greater depths than others and, in conse-quence, will gain additional protection, which might be essential for their continued existence in areas where winter conditions are particularly severe. This is thought to be so in the Canadian prairies, where *Bombus nevadensis* survives but its near relative *Bombus rufocinctus*, which hibernates more shallowly, does not.

When a young queen is mated, and her fat-bodies are well developed, she is ready to leave her maternal colony for good. Before doing so, however, she will fill her crop with honey from the colony's food stores. She then flies off in search of a suitable place to spend the winter.

Having located a likely-looking site a bumblebee will fly low over the ground, tacking gently from side to side. She may also settle at intervals to examine the surface more closely and, if conditions appear suitable, will then begin to burrow into the ground. If the soil is too hard, or if her excavations suddenly fall foul of a stone, root, or other obstruction, the queen will find somewhere else to dig. In particularly stony sites, chalky slopes with a flinty soil afford good examples, half-finished and abandoned workings are sometimes quite numerous. A bumblebee uses her jaws and legs for digging, and burrows away in very much the same fashion as a small mammal. At intervals, she will retreat a short distance up the tunnel, to force loosened soil out of the way. As the queen goes deeper, the tunnel becomes filled with soil, and once she has completely disappeared into the ground, all that can be seen on the surface is a tell-tale heap of freshly excavated soil lying immediately below the blocked-up entrance hole. However, these surface signs

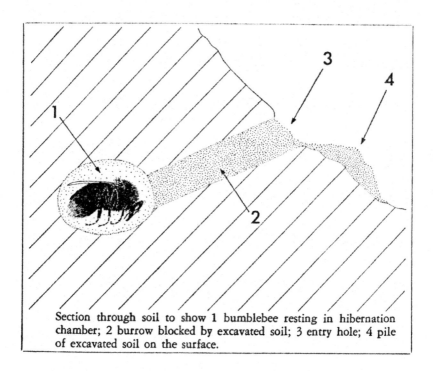

Section through soil to show 1 bumblebee resting in hibernation chamber; 2 burrow blocked by excavated soil; 3 entry hole; 4 pile of excavated soil on the surface.

do not remain obvious for long and are soon washed away by rain or destroyed by other forces. At a suitable depth, the queen hollows out a cavity about the size and shape of a walnut, in which she will spend the winter. Some bumble-bees do not bother to make a tunnel but instead hollow out a cavity in the soil immediately below a covering of leaf litter, thick moss, or other shelter. Once settled in their overwintering chambers, the queens will normally remain undisturbed until the spring.

During the construction of the hibernaculum, which may take about one or two hours to complete, the legs and, occasionally, the jaws of a bumblebee may become caked with soil. In some instances, soil particles may be found still attached to the bees following their eventual emergence from hibernation in the following year. Many years ago, before the true nature of a bumblebee colony was known, it was thought that this mud or clay was collected deliberately by the bees for building or repairing their combs.

During her long winter sleep, the food reserves in the fat-body are gradually diminished, and by the spring most of the fat will have been used up. Under very cold condi-tions, fat cannot be utilized by the body and the fat-body starch reserves are then called upon to sustain life. In her dormant state, the queen's energy requirements are compara-tively low and she does not need to drink, adequate water being made available internally from the breakdown of the fat reserves. However, whilst still active at the beginning of the hibernation period, and when finally awakened in the spring, she will need a more immediately accessible food supply, which is provided by the honey stored away in her crop. It is often thought that mild winters will favour a hibernating insect such as a bumblebee. In fact, a cold, dry season is usually more suitable as this will cut down the risk of disease breaking out and will reduce the chances of the insect being disturbed. Also, as described above, cold as such can usually be tolerated without difficulty and will ensure that the valuable food reserves available to the hibernating insect are used economically.

Bumblebees are aroused from their winter slumber by the warmth of the spring but, as previously mentioned, they do not all appear at the same time. It may be that certain species will become active at higher temperatures than others. However, the conditions under which a particular species hibernates also influence the time of emergence. For example, the earliest species to appear in southern England, *Bombus pratorum*, *Bombus lucorum* and *Bombus terrestris*, all overwinter in places that warm up more quickly in the spring than those chosen by later-emerging bumblebees, such as *Bombus hortorum* and *Bombus lapidarius*. Later-appearing species also tend to hibernate at greater depths than earlier-emerging ones.

Factors other than temperature may also influence the emergence of overwintering bumblebees, but this aspect of hibernation has never been fully explained.

Chapter 6

Cuckoo bumblebees

THE ANTI-SOCIAL HABITS of cuckoos are known to almost everyone, but it will surprise a great many people to find that equally sinister activities also take place within the world of the bumblebee. In the struggle for existence, competition between individuals is an accepted fact of life and, as in other animals, rivalry is often seen amongst bumblebees. In spring, for instance, recently formed colonies are sometimes invaded by queens that are still searching for suitable homes, such hostilities most often occurring when spring queens are particularly abundant or when acceptable nest sites in a normally productive region are few and far between. Indeed, if competition in an area is fierce, a single nest may be assaulted by a succession of different queens. The occupying queen is often overthrown during an attack but, if she is the stronger individual, she may successfully defend her domicile and slay the aggressor. When workers are present to help secure the nest, the foundress queen is likely to remain victorious, and colonies are rarely subject to attack once workers are numerous. Certain *Bombus* species are more inclined to act as invaders than others, usually storming nests headed by queens of their own kind. However, some specialize in raiding colonies of other species. A good example of this occurs in Arctic Canada where *Bombus hyperboreus* queens regularly take over nests of *Bombus polaris*. In such cases, the *hyperboreus* queen is particularly cuckoo-like in habit since she usually does not then produce workers of her own, as would a 'normal' *Bombus*, but has all her eggs reared into males and queens by the surviving *polaris* workers. But even more remarkable are the bumble-bees called *Psithyrus* which, in order to survive, are obliged to raid *Bombus* nests, as they have long since lost the ability to form and look after colonies of their own. The lives of these cuckoo bumblebees make a particularly engrossing study, and there is still much to be learnt about them.

At first glance there is little difference in appearance between a true and a cuckoo bumblebee. Closer examination, however, will show that the female *Psithyrus* lacks pollen-baskets on her hind legs and that her body's armour plating is tougher, thicker and relatively inflexible. The jaws and sting are more strongly built than those of *Bombus*, and clearly adapted for fighting; the cuckoo is also less hairy than a true bumblebee, her coat being rather thin and, more often than not, of a comparatively drab appearance. These highly specialized bumblebees have lost the ability to secrete wax and are now entirely dependent upon *Bombus* for pur-poses of brood-rearing. Differences between males of *Bombus* and *Psithyrus*, however, are slight (presumably since both kinds follow similar life styles) and likely to pass unnoticed by the casual observer.

Female cuckoo bumblebees spend the winter in similar places to *Bombus* queens, but they make their appearance somewhat later in the spring, at a time when many true bumblebees are searching for nest sites or are already establishing their colonies. Much of their time in spring is spent resting on flowers, where they sip nectar and also feed on pollen. Cuckoo bumblebees move about awkwardly and are very lethargic; they so clearly lack the busy, industrious nature of the true bumblebees. When lazing on flowers, such as dandelions (one of their favourite food sources), they frequently become dusted with pollen grains which, unlike *Bombus* they are ill-equipped to remove. Occasionally, specimens are seen that are so covered in pollen that it is at first difficult, if not impossible, to recognize the species! As in the case of *Bombus*, nectar and pollen are essential food materials for these overwintered cuckoos and soon their ovaries begin to develop. This change within them also marks a change in their behaviour, and instead of whiling away their time on flowers or elsewhere, they begin their search for a suitable home. In favourable weather they are then to be seen flying low over the ground, exploring likely-looking spots for evidence of bumblebee nests. The flight of a female *Psithyrus* is rather laboured, while the hum has a deeper, softer tone than *Bombus*; with experience, it is usually possible to distinguish an active female cuckoo bumblebee without difficulty.

In locating a nest, the cuckoo bumblebee is no doubt guided initially and attracted by the same signs that might appeal to a *Bombus* queen, but over a short distance she is able to trace a colony by smell. It was once believed that nests deep in the ground were unlikely to fall victim to the attentions of cuckoo bumblebees, but it is now known that this is not necessarly so, and records exist of *Psithyrus*-ridden colonies located several metres from the surface. Colonies sited high above the ground, such as those in birds' nests, are also liable to attack. This indicates that, as with true bumblebees, at least some kinds of *Psithyrus* are very adaptable in their choice of a home. However, each type of *Psithyrus* will usually select only the nests of one particular species for their evil purposes, although those of a close relative of the normal victim may also be acceptable. A cuckoo bumblebee is probably able to recognize the nest of a suitable host species by the characteristic scent.

Having located the entrance to a bumblebee colony, the *Psithyrus* slowly but deliberately attempts to enter. At this point she is usually confronted by the worker bees, and if their resistance is sufficiently ferocious and persistent she may be forced to retreat. On occasions, she may even be killed and unceremoniously dragged or thrown out of the nest. Wherever possible, however, she attempts to bypass the inhabitants, feigning death if approached, and tries to hide beneath the comb or elsewhere amongst the nest material. In this way she is soon able to take on the characteristic odour of the colony, which will be of assistance in her eventually attempting to gain a permanent foothold in the nest.

It is important for the survival of *Psithyrus* that attacks are launched at the right stage of development of the host colony. The best time for the cuckoo to enter is when the first brood of *Bombus* workers is present but before the appearance of the next. These early worker bees tend to be smaller and less aggressive than those of later broods, and they are unlikely to offer more than token resistance to such a powerful adversary. If attacks by the *Psithyrus* are made too late, then she may be unable to subdue or overcome the defenders and her take-over bid will fail, but usually only after some of the workers have bravely lost their lives. On the other hand, if her foray on a nest is too early she may be opposed directly by the foundress queen who, at this stage, is at her most aggressive. Here again, a fierce battle will commence. Both bees will lock in a deadly embrace, each intent upon stinging the other. Bumblebees are particularly vulnerable in the neck region, and in the flexible areas between the hard plates of the abdomen, but since the normally stronger *Psithyrus* is better protected she has a considerable, and usually unassailable, advantage. Sooner or later the sting of the cuckoo will find its mark and the defending *Bombus* queen will be killed. Unfortunately for the victorious cuckoo, however, a very young colony cannot then become strong enough in workers for *Psithyrus* brood to be reared; it may even be left totally workerless, in which case it would have to be abandoned. However, in most cases, *Psithyrus* times her attack on a nest to perfection and enters a colony of adequate strength without causing a major uprising.

Once she has become successfully installed in a colony, the cuckoo bumblebee will seek out and destroy any *Bombus* eggs. She will also break down clumps of young larvae. She usually tolerates older brood clumps since completion of their development will not overburden the economy of the colony and will soon provide additional worker bees to serve her. Any belligerence on the part of the existing workers is quickly dealt with, the offenders each being briefly grasped, threatened and chastised but not stung. These bees will be required to assist her in rearing the forthcoming *Psithyrus* brood, and killing them would only deplete their numbers and serve no useful purpose. During this period, the *Bombus* queen, if still alive, usually keeps out of the way of the cuckoo, although she may at first be bold enough to mount a tentative and half-hearted attack. So long as the queen keeps her distance and does not attempt to lay eggs, the *Psithyrus* too shows no sign of wishing to provoke a fight, and both bees will appear to live together in apparent harmony. However, as time passes, the *Bombus* queen becomes increasingly listless and almost mournful of her impending doom. Details of the relationship between a cuckoo bumblebee and its host vary somewhat from species to species, but sooner or later the *Bombus* queen is usually killed, frequently at about the time that the *Psithyrus* female lays her first eggs.

The cuckoo bumblebee builds her own egg cells but as she has no wax-producing glands she must first collect wax from the *Bombus* comb. The cells are constructed in similar positions to those of *Bombus*, but are rather crude and thicker-walled. *Psithyrus* eggs are relatively long and thin, and the female may lay some twenty to thirty in a single cell, the whole procedure taking no more than a few minutes. This high level of productivity, compared with *Bombus*, is aided by the greater number of ovarioles developed in a *Psithyrus* female, there often being from ten to fifteen in each ovary. At this stage, certain of the more truculent workers may attempt to break down the cell and destroy the *Psithyrus* eggs, but the cuckoo is usually well able to defend them, forcing the bees into submission. Once the eggs hatch, the larvae are fed and tended by some of the *Bombus* workers, while other bees continue to fly into the field to collect further supplies of nectar and pollen. When a colony is taken over by *Psithyrus* several of the *Bombus* workers will eventually lay eggs. The cuckoo is always wise to their activities, however, and diligently breaks down the cells and eats the contents, so that this clandestine *Bombus* brood never develops.

Growth of a *Psithyrus* brood clump follows a similar course to that already described for *Bombus* and, eventually, when feeding is completed, familiar parchment-like pupal cocoons appear. *Psithyrus* brood batches will produce males, or females equivalent to queens, but not workers. The number of new cuckoos reared in a colony varies considerably and will depend very much upon the efficiency and size of the slave-like *Bombus* worker force. There are generally fewer workers in colonies headed by *Psithyrus* than in successful, non-parasitized ones, so that such colonies tend to be less prolific in terms of their sexual brood-rearing potential. Also, a *Psithyrus*-ridden colony will decline more rapidly and die out earlier in the season than normal.

Psithyrus males fly away from their maternal nest a few days after they are produced, and may then be found visiting or resting on flowers, often in company with males of true bumblebees. They may also be seen following their unending flight-paths in search of a mate. The young females, however, are far less often observed. As in the case of *Bombus*, once they have mated and built up adequate food reserves in their bodies, they will immediately seek out their winter quarters and disappear until the following spring.

Chapter 7

Other enemies and nest associates

ADULT BUMBLEBEES fall victim to many parasites and predators. They are also subject to infection by bacteria, viruses and other micro-organisms; these include the protozoan parasite *Nosema bombi*. This is virtually identical to *Nosema apis* which causes the well-known dysentery disease of honey bees.

One of the most remarkable relationships is that between bumblebees and a parasitic nematode called *Sphaerularia bombi*. In some areas this tiny, worm-like creature is a major enemy of bumblebees and, in spring, can destroy large numbers of the overwintered queens. This frightening pest is also harmful to cuckoo bumblebees. *Sphaerularia bombi* spends much of its life inside bumblebees, which are invaded by one or more pregnant females during the late summer and autumn, whilst resting in their hibernation chambers. At the time of attack, each worm is like a small, whitish thread no more than 2 mm long. However, once inside a host, the parasite's uterus everts and soon grows into a large, sausage-like sack 10 to 20 mm in length, while the rest of the body remains attached as a minute appendage, once thought to be the male parasite enjoying a permanent state of copulation! By the time that the host emerges from hibernation, eggs are already developing on the parasite uterine wall and, later in the spring, they are set free into the body cavity of the bee. Shortly afterwards, tiny juveniles hatch from the eggs, forming a milky cloud in the host's blood. It has been estimated that as many as 100,000 individuals may occur in a single parasitized bumblebee. The presence of the parasite prevents the bumblebee's ovaries from developing by upsetting the hormonal balance of the body. The host's behaviour is also affected and, instead of searching for a nesting place, she will seek out and frequent old hibernation sites, behaving much like a young queen attempting to find somewhere to overwinter. These parasitized bees become more and more lethargic as the spring progresses, their rather lazy flight and abnormal behaviour at once distinguishing them from healthy individuals. Whilst the host is scratching about and flying over a hibernation site, and perhaps sometimes dying there, young parasites will escape from her body and enter the soil. Some will inevitably quit their host in unsuitable places but, because of the rather specific nature of bumblebee hibernation sites, many will find themselves congregating in ideal situations where later in the year, after they have matured and mated, females can locate new hosts and once again commence the parasitic life-cycle.

Foraging bumblebees are often parasitized by insects, including conopid flies. These wasp-like creatures have elongate, and often colourful, abdomens and large heads; they are also known as 'big-headed flies'. Conopid flies occur in early summer, when they frequent flowers to await the arrival of suitable host bumblebees. When a bumblebee approaches, or alights on a flower, the female conopid pounces and briefly grasps her prize whilst injecting an egg into the abdomen. The luckless bumblebee is then released and flies away, unaware of her fate. The parasite egg soon hatches into a white, pear-shaped grub which anchors itself to a tracheole and begins to feed on the host's body tissue. The bumblebee is not affected immediately, and will continue to forage. As the parasite grows, however, the host gradually becomes weaker and will eventually die. Shortly afterwards, the parasite com-

c*

pletes its nourishment and then pupates within a protective, barrel-like puparium formed from the cast-off larval skin, still housed in the now more-or-less shell-like remains of the host's abdomen. The parasite overwinters within the puparium and finally emerges in the following summer. Parasitized worker bumblebees often die in the nest and their swollen, rather than shrunken, abdomens immediately label them as conopid fly victims.

Bumblebees also suffer from a form of acarine disease, caused by a curious mite called *Bombacarus buchneri*, which lives in the frontal air-sacs in the abdomen. The female parasites mature into bloated, balloon-like animals about 0.5 mm across, each containing numerous eggs. These eggs are eventually released into the air-sac, to form clusters of up to fifty, which then hatch into minute adults. After mating, the young females either develop to maturity in the same host or escape through the abdominal spiracles and eventually enter new hosts. Attacks of this parasite tend to be unimportant but heavily infested bumblebees, which tend to occur later in the season, are noticeably weakened.

Adult bumblebees are preyed upon by asilid or robber flies, dragonflies, spiders and certain vertebrates. Although birds, such as great tits, sometimes attack drunken bumblebees foraging on lime trees, in more normal circumstances they tend to avoid them. However, shrikes (also known as butcher-birds) are a notable exception. They will frequently capture bumblebees and impale them on thorns in their larders.

During the course of the summer a bumblebee nest becomes the home of a wide range of animals. Some of these, including many mites, beetles and fly maggots, perform a useful function by helping to dispose of waste matter and debris from the comb, thereby preventing the outbreak of disease. Several such creatures are specialized scavengers in bumblebee nests and have developed a close relationship with their hosts. Recently-overwintered bumblebee queens are frequently seen with numerous light brown, crab-like mites called *Parasitus fucorum* clustered between their thorax and abdomen. These passengers are not parasites, as their name would suggest, but usually live and breed in bumblebee nests where they feed on faeces, scraps of pollen and other debris. Their presence on adult bumblebees is merely their way of hitching a lift from the colonies of one season to those of the next and, in the spring, they will disembark from their transporter as soon as a nest has been established. Various other mites, including the abundant but minute and scale-like *Kutzinia laevis*, are also phoretic on bumblebees for the same reason. *Kutzinia laevis* feeds on moulds in bumblebee colonies, as do less specialized visitors such as the white, long-haired scavengers *Glycyphagus domesticus* and *Glycyphagus ornatus*. The complex mite fauna in a bumblebee nest also includes an array of predatory species.

Three insect scavengers in bumblebee nests are also worthy of mention, the beetle *Antherophagus nigricornis* and the flies *Fannia canicularis* and *Volucella bombylans*. *Antherophagus* beetles occur commonly on flowers in May and June, and when a foraging bumblebee comes along they will try to grab hold of the tongue, a leg or an antenna. If a beetle is successful in gaining a hold, it is then carried back to the nest where it drops below the comb to begin breeding. The yellowish beetle larvae are very common in late summer, when their activities greatly assist in the final breakdown of the disused comb. *Fannia* adults are similar to house flies in general appearance and they probably locate bumblebee nests by smell. Their larvae are greyish-brown, adorned by distinctive spines which run along the back and sides of the body. These larvae feed on faeces at the base of the comb and are especially numerous in any damp, mucky corners below the nest material. *Volucella bombylans* adults look very much like small bumblebees, and are often mistaken for them by the casual observer. One colour form is black and yellow with a white tail, and is not unlike *Bombus jonellus* or *Bombus hortorum*; another is black with a red tail, mimicking *Bombus lapidarius* or *Bombus ruderarius*. The female fly enters a bumblebee nest during the summer and immediately lays numerous clusters of elongate eggs, which are stuck to the comb or nest material by a sticky, protective coating that soon hardens. When attempting to gain access to a colony, *Volucella* is often attacked by the guard bees. However, even if she is killed, her eggs will still be laid by the reflex action of her ovipositor. The larvae are larger than those of *Fannia*, and have six prominent spines at the hind end of the body. Although normally scavengers, feeding below the comb, they may occasionally attack the brood, particularly towards the end of the season. Numerous small, parasitic wasps occur in bumblebee nests and some of these attack the larvae of these scavengers. *Antherophagus* and *Fannia* larvae, for example, are frequently parasitized by *Blacus paganus* and *Stilpnus gagates* respectively.

There are two major parasites of bumblebee brood, the fly *Brachicoma devia* and the velvet-ant *Mutilla europaea*. *Brachicoma* looks very much like a house fly, but is larger than *Fannia*, and invades bumblebee nests during the spring. The female is viviparous, that is, instead of laying eggs, she gives birth to small, live maggots, placing them in the brood cells of the host. These whitish, parasitic grubs remain quiescent until the bumblebee larvae spin their final cocoons and enter the pre-pupal stage. They then attack, often up to four feeding on a single host. They drain off the body fluid and, within a few days, will reduce the pre-pupa to an empty sack. Each parasite then escapes from the host cocoon and worms its way into the debris of nest material at the base of the comb. Here it pupates within the hardened remains of its own skin. The young flies emerge a week or so later and, after mating, another attack is mounted on the bumblebee brood. *Brachicoma* can be very harmful and infested colonies, which are characterized by an unpleasant smell emanating from the parasitized comb, may die out prematurely. There are several parasite generations in a season, the final one overwintering as pupae in or near the old host nest.

Bumblebee brood suffers an equally gruesome fate at the hands of *Mutilla europaea*. The adult of this wasp is an unusual, ant-like and handsome insect with a black head, a reddish thorax and a black, hairy abdomen blazoned with two or three silvery crossbands. The female, unlike the male, is completely wingless. She enters bumblebee nests in the early summer and deposits eggs singly in young pupal or pre-pupal cocoons. On hatching, each grub begins feeding on the host which, by the end of the feeding period, may be completely devoured. The parasite then spins a tough cocoon of its own, and the perfect adult insect emerges a week or so later. The adult mutillid is able to produce a characteristic sound by moving the abdomen, and often heralds its presence (even before emerging from the cocoon), with a distinctive, hissing call. Fortunately for our bumblebees, if not for the enquiring naturalist, this fascinating and highly specialized enemy is now an uncommon insect and rarely seen.

Wax-moths are also serious enemies of bumblebees, and in some areas they cause many colonies to die out before young queens can be produced. The most important species is *Aphomia sociella*, which is particularly damaging to surface-nesting bumblebees such as *Bombus ruderarius*. Wax-moths invade bumblebee nests in July or August, probably locating them by smell. They lay up to a hundred eggs, which hatch in about a week. The wax-moth caterpillars then begin to feed on the comb, riddling it with silken tunnels, into which they retreat if disturbed. As the caterpillars grow they rapidly break down the empty cocoons and later turn their attention to the food stores and brood cells. The adult bumblebees are incapable of defending their domain against this onslaught, and the colony is soon totally destroyed. Full-grown wax-moth caterpillars are about 30 mm long and greyish or yellowish green in colour. At this stage they all crawl out of the nest and each spins a tough silken cocoon in which to overwinter and, finally, pupate. These cocoons are often formed in a ball-like mass under a large stone or other shelter near the nest entrance.

Finally, whole colonies are sometimes wiped out by mammals, such as badgers and (in America) skunks, which eat the comb, the brood and the adult bees. Early in the season, many young colonies are destroyed by mice, shrews and voles but, unlike larger animals, they can usually only mount a successful incursion before the appearance of the workers and whilst the foundress queen is out foraging. The damage caused by small mammals, however, is more than countered by their usefulness in providing valuable nesting places for bumblebees.

Chapter 8

Bumblebees and man

MAN WAGES a constant war against insects. In England and Wales alone, several hundred tonnes of pesticides, costing millions of pounds, are applied to agricultural crops each year, and world figures are astronomical. Various insects destroy crops, damage buildings and their contents, attack domesticated animals, or are responsible for transmitting diseases. But the adage that 'the only good insect is a dead one' is far from being true. Insects play a major role in maintaining the balance of nature and many, such as butterflies, have a strong aesthetic appeal which often brings out man's best intentions on the side of wildlife. Insects are also a vital link in the food chain of many animals; our insectivorous birds, for instance, which bring such pleasure to man, could not survive without them. In addition, large numbers of insects can rightfully be described as directly beneficial to man. These include numerous parasites and predators of pests; the many kinds of parasitic, ichneumonid wasps, predatory beetles such as ladybirds, and hoverfly larvae, are familiar examples. Man also benefits from bees and other insects that regularly pollinate his flowers and crops. It is in this context that the bumblebee is particularly useful.

Bumblebees are efficient pollinators because their large size and hairy coats make them ideal for picking up and transferring pollen grains from one flower to the next. Also, their specialization and constancy to particular plant species during their foraging trips increases the chances of the correct type of pollen being passed from flower to flower. Bumblebees, in common with other bees with larvae to feed or brood cells to provision, have another major advantage in that they visit many more flowers and forage more consistently than insects that have only to obtain nourishment to sustain themselves. In this context, it follows that *Bombus* is more helpful than *Psithyrus* and that a worker bumblebee is more effective than a male.

Since bumblebees work longer hours and in poorer conditions than most other insects, they are greatly favoured as pollinators in certain regions and seasons. For example, they are very useful to fruit growers during the spring when the weather may be unfavourable for other bees, including honey bees. Bumblebees are often present in commercial and garden orchards at blossom time, particularly where surrounding countryside is favourable for their survival; they may also be seen in bush-fruit plantations, including blackcurrant holdings. Later in the season, bumblebees are also abundant on raspberry flowers, for which *Bombus pratorum* has a particular liking. Long-tongued bumblebees, such as *Bombus hortorum*, are important wild pollinators of lucerne, red clover, field bean, runner bean, and certain other crops. A major disadvantage, however, is that numbers are often too low for adequate pollination and, since the size of populations varies so much from year to year, bumblebees alone cannot be relied upon.

It is difficult to quantify the value of bumblebees to agriculture and horticulture since many flowering crops can set fruit or seed without them. Nevertheless, certain plants, including important crops like clover, field and runner beans, undoubtedly need to be pollinated by insects, such as bumblebees, before they can be adequately fertilized. Even where this is not the case, yields and fruit or seed quality are often improved by insect visitation. Also, pollination by insects can lead to earlier and more even ripening of crops, with obvious advantages for mass harvesting. The value of long-tongued bumblebees to red clover seed production was clearly demonstrated in New Zealand when, following their introduction from Britain towards the end of the last century and their successful establishment, yields increased dramatically.

Bumblebees can also play a useful role as pollinators in plant-breeding and experimental work. Either whole colonies may be placed in suitable enclosures and the bees allowed to forage on the plants to be pollinated or, alternatively, males, workers or queens can be employed separately. Over-wintered queens are particularly helpful in the production of brassica seed under controlled conditions, but to prevent unnecessary depletion of natural bumblebee populations it is desirable, if at all possible, to use individuals parasitized by the nematode *Sphaerularia bombi.* This technique has proved successful in Holland, but enough parasitized queens are often hard to come by.

A recent survey of the British bumblebee fauna has indicated that some of our species, including *Bombus humilis,* *Bombus ruderatus* and *Bombus sylvarum,* are more restricted in their distribution than formerly and many are now only numerous in unspoilt countryside offering a wide range of wild flowers. Chalk downland habitats, for example, still maintain a varied bumblebee fauna, whereas the open farm-land regions of eastern England have become relatively impoverished. Large areas of crops such as beans or clover can be more than useful food sources during their flowering periods but bumblebees and other wild pollinators are unlikely to be abundant unless their requirements for over-wintering, nesting, and foraging whilst the crops are not in flower, are also met. In general, diversity in the environment is particularly favourable for bumblebees, although a few species, such as *Bombus jonellus* and *Bombus lapponicus,* have special habitat requirements that will inevitably restrict their distribution to certain parts of the country (see Appendix).

Pesticides are often blamed for the decline in wild bee populations and, undoubtedly, many bumblebees are killed by chemical sprays each year. However, losses of forage plants by the widespread use of herbicides, as an inevitable concession to modern farming practice, are even more detrimental. Such depredations are compounded when sprays are also applied to waste and uncultivated land or when such areas are unnecessarily cut just before or during the flowering period. Fire can also destroy the foraging and nesting potential of banks, hedgerows and verges, rendering them useless to bumblebees and many other forms of wildlife for months, if not years. It is regrettable that such destruction is often due to carelessness and usually carried out for no good reason. The major factor in reducing bumblebee numbers, however, is modern land use which so often demands the widespread and usually permanent destruction of their haunts. However, the influence of man is not always detrimental. Some bumblebee species, including *Bombus pratorum*, are much attracted to man-made habitats such as parks and gardens. Indeed, many of our most common species are capable of occupying and surviving in suburban and urban areas.

Bumblebees can be encouraged to survive, and perhaps even to multiply, by careful management of the environment, such as the conservation or provision of suitable flowers, trees and shrubs. These include dead-nettles, woundworts, knapweeds, flowering cherries, sallows, and many others. Particularly in North America, attempts have been made to increase bumblebee populations by providing artificial nest-boxes in the field for occupation by the overwintered queens, and such techniques have also been tried with some success in Europe. It is also possible to persuade bumblebee queens of some species to establish colonies in confinement by providing them with suitable food and brood-nest conditions. However, these techniques are by no means perfected and the bumblebee is a long way from being regarded as a domesticated animal.

Bee products, such as beeswax and honey from honey bee colonies, are also of value to man, and it is sometimes wondered whether bumblebees could prove useful in this respect. However, the amount of wax produced in a bumble-bee colony, even by the most prolific producers such as *Bombus lapidarius*, is insufficient for worthwhile exploitation and the relatively small quantity of honey stored in a bumble-bee comb is but small recompense for the trouble needed to obtain it. In addition, unlike honey bee colonies which can be farmed without adverse effects, plunderous interference of bumblebee nests will almost invariably harm their overall economy and cannot be justified. Nevertheless, captive bumblebees can serve an educational function. Whole colonies may be placed in nest-boxes and their intricate social behaviour and organization investigated to the full. Bumble-bees can also be observed in the wild, their nesting, mating and foraging habits, for example, making ideal subjects for local study. There is also much to be learnt about their parasites and nest commensals. Any intending researcher, however, should ensure that wherever possible bumblebees are correctly identified as to species, if necessary seeking expert help or guidance, as the behaviour and physiology of these insects varies considerably from species to species and what is true of one may not necessarily apply to another. Observations where the various species have not been named, therefore, are often of little scientific value.

Mention has been made of the aggressive behaviour of certain bumblebees when defending their nests. As far as man is concerned, however, bumblebees are to be respected rather than feared. Occasionally, he may be stung, say when interfering with a bumblebee nest or when disturbing foragers engaged in their lawful pursuits; however, it is then entirely his own fault! Bumblebee colonies established in gardens or outbuildings sometimes cause concern, particularly if small children are about, but destruction of such nests should be a last resort. Normally, colonies are only discovered when, after many weeks, they reach the height of their development and the passage of numerous workers in and out of the nest attracts attention. By then, they will normally have only a few more weeks of active life before their ultimate decline and will be at the most vital period of their development, involving the all-important stage of queen production upon which the future of the species depends. Bumblebee nests are often destroyed out of ignorance, as it is not always appreciated that they are useful insects and that occupation of a nesting place is only a temporary one. Most bumblebees found nesting in gardens are comparatively mild-tempered, even when defending their nests, and they are unlikely to be a problem. Indeed, a bumblebee colony should cause no more inconvenience than a birds' nest and, in the interests of wildlife conservation, and bumblebees in particular, wherever possible it should be allowed to complete its natural cycle unmolested. Many people go to great lengths to encourage birds to visit and breed in their gardens or estates. The bumblebee deserves similar support as it is both beneficial to man and one of nature's most endearing and enigmatic insects.

Appendix I

List of the British species

Genus Bombus (Latreille) – TRUE BUMBLEBEES

1 *B. soroeensis* (Fabricius) ⎫
2 *B. lucorum* (Linnaeus)   ⎪
3 *B. magnus* (Krüger)   ⎪
4 *B. terrestris* (Linnaeus) ⎪
5 *B. cullumanus* (Kirby)   ⎬ 'pollen-storers'
6 *B. jonellus* (Kirby)   ⎪
7 *B. lapponicus* (Fabricius) ⎪
8 *B. pratorum* (Linnaeus)   ⎪
9 *B. lapidarius* (Linnaeus) ⎭

10 *B. hortorum* (Linnaeus)   ⎫
11 *B. ruderatus* (Fabricius)   ⎪
12 *B. humilis* (Illiger) =   ⎪
    helferanus (Seidl)   ⎪
13 *B. muscorum* (Linnaeus)   ⎪
14 *B. pascuorum* (Scopoli) =   ⎪
    agrorum (Fabricius)   ⎬ 'pocket-makers'
15 *B. ruderarius* (Müller) =   ⎪
    derhamellus (Kirby)   ⎪
16 *B. sylvarum* (Linnaeus)   ⎪
17 *B. distinguendus* (Morawitz)   ⎪
18 *B. subterraneus* (Linnaeus)   ⎪
19 *B. pomorum* (Panzer)   ⎭

Genus *Psithyrus* (Lepeletier) – CUCKOO BUMBLEBEES
20 P. *bohemicus* (Seidl)
21 P *vestalis* (Geoffroy in Fourcroy)
22 P. *rupestris* (Fabricius)
23 P. *barbutellus* (Kirby)
24 P. *campestris* (Panzer)
25 P. *sylvestris* (Lepeletier)

Appendix II

Notes on the British Species

Identification of the various species requires a certain degree of expertise and should be based on features other than colour patterns. However, this subject is beyond the scope of the present book. Full descriptions and illustrated keys to the British species are included in the monograph *Bumble-bees* (Alford, 1975). The following descriptions, based on females, are necessarily brief and are intended merely as guides; in many cases they are not strictly applicable to males.

The species likely to be encountered by the average reader are marked with an asterisk.

1 *Bombus soroeensis.* Small-sized; thorax black with yellow collar, abdomen black with yellow crossband and white tail. Widely distributed in England, Scotland and Wales, but very local and usually rare; absent from Ireland.

*2 *Bombus lucorum.* Medium-sized; thorax black with yellow collar, abdomen black with yellow crossband and white tail. Widely distributed and abundant almost every-where.

3 *Bombus magnus.* Large-sized; very similar in appearance to *lucorum.* Mainly found in the west and north of Britain and in Ireland, where it is sometimes abundant.

*4 *Bombus terrestris.* Large-sized; thorax black with yellowish collar, abdomen black with yellow crossband and buff tail. Workers are coloured more like *lucorum.* Absent from the extreme north of Scotland but otherwise widely distributed and usually common, especially in southern England.

5 *Bombus cullumanus.* Formerly recorded from a few chalk-land localities in England, but probably now extinct.

6 *Bombus jonellus.* Small-sized; thorax black with yellow collar and scutellum, abdomen black with yellow crossband and white tail. A heathland and moorland species; local in England but more generally distributed in Ireland, Scotland and Wales.

7 *Bombus lapponicus.* Medium-sized; thorax black with yellow collar and scutellum (usually), abdomen mostly orange with black base. Mainly a local inhabitant of mountains and moorlands; associated with bilberry. Until recently, unknown in Ireland.

*8 *Bombus pratorum.* Small-sized; thorax black with yellow collar, abdomen black with yellow crossband and orange-red tail. Widely distributed and often abundant.

*9 *Bombus lapidarius.* Large-sized; thorax black, abdomen black with red tail; corbicular hairs black. Generally distributed and often common.

*10 *Bombus hortorum.* Medium-sized; thorax black with yellow collar and scutellum, abdomen black with yellow crossband and white tail. One of our most widely distributed and common species.

11 *Bombus ruderatus.* Large-sized; similar to *hortorum* but usually less brightly coloured – a completely black form occurs. Mainly found in the southern half of England but it is local and generally uncommon; absent from Ireland and Scotland.

12 *Bombus humilis.* Small-sized; thorax orange-brown with a few black hairs near wing bases, abdomen yellowish brown with brown crossband. Locally distributed in southern England and Wales, but absent from Ireland and Scotland.

13 *Bombus muscorum.* Medium-sized; thorax orange-brown, abdomen yellowish brown. Widely distributed in northern England, Ireland, Scotland and Wales, producing various distinct colour forms; certain off-shore island races are very striking in appearance. In southern England it is local and mainly found in marshes and coastal habitats.

*14 *Bombus pascuorum.* Small-sized; thorax and abdomen yellowish or reddish brown with black hairs intermixed and often numerous; an extremely variable species. Abundant almost everywhere.

*15 *Bombus ruderarius.* Small-sized; thorax black, abdomen black with rusty-red tail; corbicular hairs reddish. Widely distributed and often common, especially in southern England, but scarce or absent in many areas.

16 *Bombus sylvarum.* Small-sized; thorax pale greenish or yellowish grey with indistinct black interalar band, abdomen pale greenish or yellowish grey marked with black and an orange tail. Widely distributed, particularly in the south of England, but generally uncommon; rare in Ireland and absent from Scotland.

17 *Bombus distinguendus.* Large-sized; body yellow or brownish yellow with back interalar band. Generally uncommon; most frequently recorded in the north and west of Britain, and in Ireland.

18 *Bombus subterraneus.* Large-sized; thorax black with brownish yellow collar and scutellum, abdomen black with narrow, brownish white crossbands and a whitish tail; a very short-haired species. Widely distributed in southern England but usually rare; absent from both Ireland and Scotland.

19 *Bombus pomorum.* The only British specimens are from Kent; last recorded in 1864 and now, no doubt, extinct.

20 *Psithyrus bohemicus.* Large-sized; thorax mainly black with yellow collar, abdomen black with a few lemon-yellow hairs immediately in front of the white tail. Widespread and often common, except in the midland, southern and eastern counties of England. Attacks colonies of *Bombus lucorum*.

21 *Psithyrus vestalis.* Large-sized; similar in appearance to *bohemicus* but with yellower hairs immediately in front of the white tail. Most common in southern England, where it victimizes *Bombus terrestris*; absent from Ireland and Scotland.

22 *Psithprus rupestris.* Large-sized; body black with red tail; wings distinctly brown. Today, our least common cuckoo bumblebee, most frequently reported from southern England and Ireland. An enemy of *Bombus lapidarius*.

23 *Psithyrus barbutellus.* Medium-sized; thorax black with brownish yellow collar and scutellum, abdomen black with yellowish crossband and white tail. Widely distributed and by no means uncommon. An enemy of *Bombus hortorum*.

24 *Psithyrus campestris.* Medium-sized; similar in appearance to *barbutellus*, but breeding in colonies of *Bombus pascuorum* and, perhaps, *Bombus humilis*. An all black form occurs widely and, in Scotland, there is an almost entirely yellowish-coloured sub-species. Widely distributed and sometimes relatively common.

25 *Psithyrus sylvestris.* Small-sized; body mainly black with yellow collar and a mainly white tail. Widely distributed and often common. Attacks *Bombus pratorum* colonies; at least in some areas, *Bombus jonellus* is also a host.

Further Reading

ALFORD, D. V. (1975): *Bumblebees.* Davis-Poynter, London

FREE, J. B. (1970): *Insect pollination of crops.* Academic Press, London and New York

FREE, J. B. & BUTLER, C. G. (1959): *Bumblebees.* Collins, London

PLATH, O. E. (1934): *Bumblebees and their ways.* Macmillan, New York

PROCTOR, M. & YEO, P. (1973): *The pollination of flowers.* Collins, London

RICHARDS, O. W. (1953): *The social insects.* Macdonald, London

SLADEN, F. W. L. (1912): *The Humble-bee, its life history and how to domesticate it.* Macmillan, London

Glossary

Abdomen : hind part of body separated from the thorax by a constriction or 'waist'.

Antennae : the pair of sensory feelers on the head of an insect.

Arthropod : member of the Arthropoda, the largest phylum in the animal kingdom; typified by their hard exoskeleton and jointed limbs.

Beeswax : wax secreted by bees (usually honey bees) and used for constructing the comb.

Caste : in social insects, a functionally and often structurally specialized individual; in bumblebees (*Bombus*), the female sex is composed of two castes, queen and worker.

Coat : the hairs covering the body constitute the 'coat' of a bumblebee. Apart from parts of the limbs, which may be more or less brownish, the body surface of a bumblebee is black. Descriptions of colour, therefore, refer to the overall appearance of the body hairs.

Collar : the pale anterior thoracic band of hairs, present in several species.

Commensals : animals living non-parasitically in association (eg in the same nest) with a member or members of another species, usually to their mutual benefit.

Corbicular hairs : the long hairs of the corbiculum.

Corbiculum : the pollen-carrying apparatus or 'pollen-basket' on the hind leg of a *Bombus* female and honey bee worker.

Corolla : the petals of a flower, joined at the base, and often forming a distinct tube.

Enzymes: catalysts produced by a living organism that promote the breakdown of certain chemical substances without themselves being altered or destroyed.

Hiberation: quiescent state in which many animals pass the winter.

Honey bee: the domesticated hive bee, *Apis mellifera*.

Honeydew: dilute sugary solution exuded by many sap-feeding insects such as aphids.

Ichneumonid wasp: one of many parasitic wasps (Family Ichneumonidae) whose hosts are often insects.

Insectivorous: insect-eating.

Instar: growth stage of an insect between moults.

Interalar band: a black band of hairs across the thorax between the wings.

Mass-provisioning: as in solitary bees, the provisioning of a brood cell with sufficient food from the outset to fulfil the requirements of a larva throughout its development, without the need for the parent to add further supplies.

Nectar: the sweet substance secreted by special glands, the nectaries, in flowers and certain leaves. Glands not in flowers are called 'extra-floral nectaries'.

Nectary: the part of a flower from which nectar is secreted.

Nematodes: non-segmented, worm-like animals in the phylum Nematoda, often called roundworms, threadworms or eelworms.

Ovariole: egg-producing tubule of insect ovary.

Parasite: an organism living in or on another (called the host) and from which it obtains food.

Phoretic: an animal practising phoresy (temporary and forcible hitch-hiking upon the body of another).

Pollen: small, powdery grains which contain the male reproductive cells of a flower. The pollen is produced in the sac-like anthers at the tips of the stamens.

Pollination: the transfer of pollen from the anthers of a

flower to the receptive, female stigma of the same flower or another flower of the same species.

Pollinator: an agent effecting pollination.

Predator: an animal that feeds by capturing and killing another (its prey).

Progressive feeding: the provisioning of a brood cell with food at intervals throughout the development of a larva.

Protozoan: single-celled animal in the phylum or sub-kingdom Protozoa.

Pupa: developmental stage of an insect during which metamorphosis (the change) from larva to adult takes place.

Puparium: protective case round a pupa, such as that formed from the cast-off larval skin of certain Diptera (flies).

Scavenger: an animal which feeds on dead organisms and debris, often performing a useful sanitary role.

Scutellum: the hind part of the thorax, often covered in pale hairs.

Social insect: one in which parent and offspring live in mutual co-operation in a common shelter or nest.

Solitary bee: one of a number of non-social species of bee.

Spermatheca: in a female, the storage organ for sperm received from the male during mating.

Spiracles: apertures (breathing pores) on each side of an insect, leading into the tracheal (breathing) system.

Tail: a term of convenience to describe the upper, hinder part of the abdomen. Descriptions of tail colour relate to its general appearance and, in detail, often exclude consideration of the hairs at the extreme tip of the abdomen which may be of a different colour.

Thorax: in an insect, the part of the body between head and abdomen, which bears the legs and wings.

Tracheoles: the fine, ultimate threads of the insect tracheal (breathing) system.

Viviparous: giving birth to live young (eg larvae) rather than eggs.

Index

CPSIA information can be obtained at www.ICGtesting.com
Printed in the USA
LVOW02s2301280815

451993LV00013B/148/P